AnswerMan Electricity for HVAC&R
A Guide to Troubleshooting
ISBN 1-930044-07-0

Printed in the United States of America

D0450408

INTRODUCTION

The AnswerMan, Electricity for HVAC&R—a guide to troubleshooting, is a practical view of what electricity is and how it works. It is more than just a textbook. Each subject is addressed by answering the most frequently asked questions about electricity.

Every effort has been made to keep it simple. The small in size format makes it easy to keep big information readily at hand.

For those of you that will be taking the HVAC Excellence Electrical Certification Exam, this manual addresses all of the subjects that are included in the test.

TABLE OF CONTENTS

WHAT IS ELECTRICITY?

Electricity is a form of energy. Other forms of energy are: heat, light, chemical (battery), atomic (power plant), and mechanical (motor). Energy cannot be destroyed, but can be converted from one form of energy to another. For example, heat is used to produce electricity which can be used to produce light. Electrical energy is the movement of electrons through a conductor.

Electrical appliances are designed to convert electrical energy to another form of energy, and thus perform useful work. Some devices produce heat, other devices produce motion or light.

WHAT IS AN ELECTRON?

All matter is made of atoms. Atoms are made up of particles called protons, neutrons and electrons. The protons and neutrons are located at the center or *nucleus* of the atom. Electrons travel in orbits around the nucleus. The protons have a positive charge. The electrons have a negative charge. Neutrons have no charge and have no effect on the electrical characteristics of the material. Electrical energy is released when electrons move from one atom to another. Electrons can be forced to pass from one atom to another. Atoms try to maintain equal numbers between positive (+) and negative (-) charges (protons vs. electrons). An atom having lost an electron becomes positively charged (+), due to the excess proton. An atom having an extra electron becomes negatively charged (-).

The Law of Electric Charges states: like charges repel and opposites attract. Excess electrons are attracted to atoms lacking electrons. To perform useful work, a constant and steady movement of electrons must be produced.

WHAT IS A POTENTIAL DIFFERENCE?

An imbalance of electrons is called a potential difference. A potential difference describes a situation where excess electrons have accumulated and are waiting an opportunity to re-connect with atoms that are lacking electrons. Therefore, electrons travel from negative atoms (-) to positive atoms (+).

There are a variety of methods used to create a potential difference (electromotive force) between two points: friction (static electricity), chemical (battery), thermoelectric (heat), photoelectric (light), and magnetic (generator).

The presence of a potential difference is sometimes called electromotive force (emf), which is further abbreviated to "E".

WHAT IS A VOLT?

The potential difference (emf) between two points can be very high or very low. The unit of measurement used to indicate the strength of the emf is the Volt.

Some typical voltages are: 1.5 volts for a flashlight cell, 12 volts for auto batteries, 120 volts for homes, 240 volts for commercial systems, etc. Voltage can vary from microvolts (millionths of a volt) to megavolts (millions of volts).

The terms potential, electromotive force (emf), and voltage mean the same thing can be used interchangeably. Most people refer to emf as volts. Remember, electromotive force is NOT electricity. It is the driving force that causes electrons to move from one atom to another.

HOW DO I MEASURE VOLTS?

Voltmeters are used to measure potential difference between two specific points. All voltage testers have two probes and the meter indicates potential difference between the two probes. Voltmeters are often used to check electrical power supply. Correct placement of the probes and understanding the readings is critical for proper troubleshooting of electrical problems. Voltmeters are available in analog or digital types. A digital meter is much easier to read because it displays the voltage directly. There is always the possibility of misreading an analog meter.

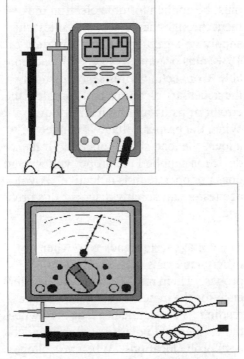

Fig. 1. Digital and Analog voltmeters

Electrical appliances are energy conversion devices (called loads) and they are designed for connection between a potential difference. A specific voltage must be applied to force electron movement through the device. When testing supply voltage, a maximum variation of 10% (plus or minus) is generally acceptable. Connecting wires serve to supply the necessary electrons and complete the circuit or pathway for electron flow. When the proper voltage is connected to a load, the load should operate. If the device is supplied the proper voltage and does not operate, it is defective. A voltage tester can quickly discover this problem.

The voltage tester reads zero when no difference exists between the two probes. There must be a potential difference for the meter to register a voltage reading. The voltmeter reads zero when the voltage and polarity are the same at both probe locations. When no voltage exists, the voltmeter reads zero. Likewise, if the voltage is 120 volts at each probe, the voltmeter reads zero. Never touch an electrical wire because a zero voltage reading was obtained. You may be reading the same potential (no difference) between the probes. Additional voltage tests are required to determine if voltage is (or is not) present. Fig. 2 shows several examples of how to use a voltmeter.

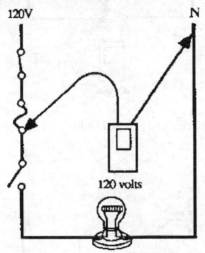

Fig. 2 (a) Voltmeter reading = 120 volts

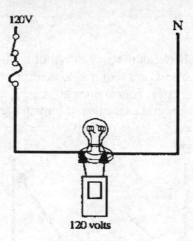

Fig. 2(b) Voltmeter reading = 120 volts

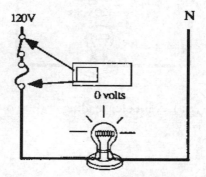

Fig. 2(c) Voltmeter reading = 0 volts

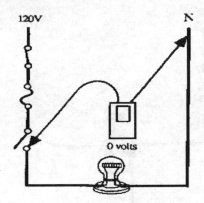

Fig. 2(d) Voltmeter reading = 0 volts.

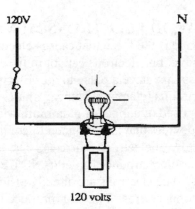

Fig. 2(e) Voltmeter reading = 120 volts.

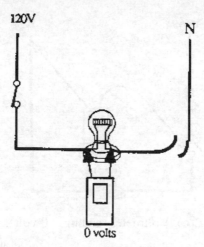

Fig. 2(f). Voltmeter reading = 0 volts.

HOW DO ELECTRONS MOVE?

Voltage is the force that causes electrons to move, but electrons cannot move unless they have a place to go. Electrical circuits (electrical pathways) are composed of copper or aluminum wires and devices that are designed to control the flow of electrons (current). The power plant produces an emf, NOT electrons. The electrons are already inside the copper wires. The emf produced by the power plant forces free electrons inside the conductor to travel to the next atom within the conductor, much like a domino effect. This electron movement (from one atom to another) occurs throughout the length of the conductor. The copper conductors provide the necessary free electrons and provide the proper pathway for electron movement.

Electricity is often compared to the flow of water because it takes the path of least resistance. However, water valves and electrical switches operate differently. Water flows when a valve is opened. Electricity flows when the switch is closed. Electrons cannot flow through an open switch or a broken wire. Any opening in the circuit (pathway) is much like a drawbridge. Switches are used in electrical circuits to act as "drawbridges" for stopping and starting the flow of electrons.

WHAT IS AMPERAGE?

The words ampere, amperage, amps, and current are frequently used to describe the quantity and intensity of electrons moving through a conductor. Amperage determines the amount of electrical conversion to another form of energy. Thus, electrical loads are "energy conversion devices". These energy conversion devices (toasters, light bulbs, motors, etc.) are used to perform useful work.

An ammeter is used to measure the quantity and intensity of electrons flowing inside a wire, or through a load. The letter "I" (for intensity) is often used to indicate amperage flow. The clamp-on ammeter is most commonly used on AC circuits and is designed to read the intensity of flow in ONE wire. The ammeter gives a false reading of zero when clamped around two wires. Fig. 3 illustrates a typical clamp-on ammeter. These are available in analog type (needle pointer) or digital readout.

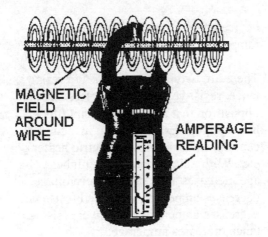

MAGNETIC
FIELD
AROUND
WIRE

AMPERAGE
READING

Fig 3. The inductive ammeter reads intensity of magnetic field around the wire and converts it to an amperage reading.

WHAT IS RESISTANCE?

Resistance refers to anything offering opposition to the flow of electrons. It is the resistance that causes energy conversion. Electron flow is energy in motion and must be controlled. The resistance controls the amount of electron flow, and thus regulates the rate at which the useful work is performed. The electrical

13

device cannot function if electron flow (amperage) is too high or too low.

There are two important types of resistance, <u>resistive</u> and <u>reactive</u>. Resistive is opposition that remains constant (does not change). Examples are: incandescent light bulbs, toaster, electric heaters, etc. With a fixed resistance; higher voltage increases amperage, lower voltage decreases amperage, increased resistance decreases amperage, and lower resistance increases amperage.

The reactive form of resistance has low resistance at start-up, but increased resistance during operation. Examples are: transformers, solenoid coils, motor windings, etc. These devices produce a magnetic field and voltage of their own that is in direct opposition to the supply voltage. This counter-voltage (counter-emf) acts as additional resistance, created only when the device is operating. Counter-emf decreases current flow after start-up and during operation of the device.

WHAT IS IMPEDANCE?

The total of both resistive and reactive opposition to ac current flow is called *impedance*.

WHAT IS OHM'S LAW?

The exact relationship between voltage (E), amperage (I), and resistance (R) was discovered by George S. Ohm. Ohm's Law is used to design electrical devices, circuits, and for troubleshooting purposes. The unit of resistance, the **ohm**, was named in his honor. The capital letter "R" is often used to indicate resistance. Another symbol for resistance is the Greek capital letter Ω (omega).

Ohm's Law is best remembered as a pie, as shown in Fig. 4. To use the pie, cover the item to be determined and follow the instructions as indicated by the horizontal or vertical lines. For example, to discover E, you must multiply I x R. To discover I, divide R into E.

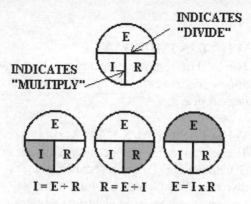

Fig. 4 Cover the unknown item and follow instructions indicated by the horizontal or vertical lines.

WHAT IS A SERIES CIRCUIT?
A series circuit has one single path for current flow. The components in the circuit are arranged so that the current must flow through the first device, then through the second device, then through the third , and so on. If the connection is broken or if one of the devices fail, current flow stops in the entire circuit.

There may be several switches in series with a motor or other device. If any one of the switches opens, current stops.

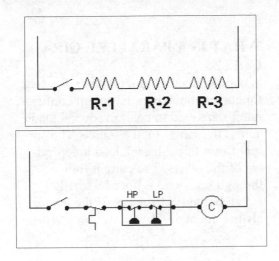

Fig. 5. Examples of series circuits.

HOW DO I DETERMINE TOTAL RESISTANCE IN A SERIES CIRCUIT?

A series circuit has only one path for current flow. Therefore the total resistance is the sum of all of the resistances in the circuit.

$R_T = R_1 + R_2 + R_3...$etc.

WHAT IS A PARALLEL CIRCUIT?

A parallel circuit has more than one path for current flow. Parallel circuit configurations are used to connect several loads across the same voltage source. The current flows through each load independent of the others. The current flow through each load is not necessarily equal, but the voltage across the load is always equal.

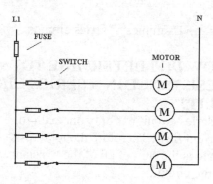

Fig. 6 Parallel circuit.

HOW DO I DETERMINE THE TOTAL RESISTANCE IN A PARALLEL CIRCUIT?

Since a parallel circuit has more than one path for current flow, adding additional paths (loads) will decrease the total resistance in the circuit.

The formula to calculate the total resistance in a parallel circuit is:

$$R_{total} = \cfrac{1}{\dfrac{1}{R_1} + \dfrac{1}{R_2} + \dfrac{1}{R_3} \ldots}$$

HOW DO I MEASURE RESISTANCE?

Ohmmeters are used to check resistance. Ohmmeters are very sensitive and measure resistance in "ohms". They measure electron movement calibrated to a voltage supplied by the meter's battery. When using an ohmmeter, supply voltage must be turned off and disconnected from the device being tested. Failure to disconnect the device from the circuit can cause severe damage to the ohmmeter, or result in false readings caused by electron flow through another circuit.

Resistance readings also reveal specific situations such as continuity, an open circuit, or a short circuit.

Continuity describes a complete path for electron flow and is revealed by zero resistance. Continuity indicates no broken wires, open switches or blown fuses.

An open circuit describes an open switch, blown fuse, broken wire, etc. Electrons cannot flow in an open circuit. The ohmmeter reveals an open circuit by unlimited resistance (infinity), or extremely high resistance (megaohms).

A short circuit is a complete circuit (continuity) where none should exist. There is very little, or no resistance in a short circuit. The ohmmeter detects a short circuit by indicating zero resistance between two points that should read extremely high resistance, or infinity.

WHAT IS A WATT?

James Watt (1763-1819) discovered the method for measuring electrical power. Electrical power is the rate at which electricity is used to perform useful work. This work is measured in units called watts. Watts are calculated by multiplying amperage x voltage: $W = I \times E$. 746 watts is equal to 1 horsepower. A wattmeter is normally located at the power entry to a building and measures the number of kilowatts (1000 watts) used per hour.

WHAT ARE CONDUCTORS?

Electrical wires are used as conductors to provide the necessary free electrons and serve as a pathway for electron flow. These wires are used to connect devices and switches to complete an electrical circuit. Silver, copper and aluminum are good conductors because they have high conductivity and low resistance to current flow. In general, any material that has 3 or less electrons in its outer orbit, called the valance ring, is considered a good conductor. Copper is the most commonly used conductor. Copper has a single electron in its valance ring.

The amount of current a conductor can safely carry without becoming over-heated is limited. This current-carrying ability is called ampacity. The ampacity of a conductor depends upon the wire's size (AWG), length, location, type and quality of insulation. The chart shown in Fig. 7 is a partial list of wire sizes, resistance, and ampacity for standard sizes of copper and aluminum wires.

Gauge No. (AWG)	OHMS per 1000 FEET	Ampacity Copper	Ampacity Aluminum
0000	0.050	230	180
000	0.062	200	155
00	0.080	175	135
0	0.100	150	120
1	0.127	130	100
2	0.159	115	90
3	0.202	100	75
4	0.254	85	65
6	0.40	65	50
8	0.645	50	40
10	1.02	30*	25
12	1.62	20*	18
14	2.57	15*	
16	4.10	10*	
18	6.51	5*	

Fig.7. Ampacity of Commercial Wire
*Load current rating and overcurrent protection shall NOT exceed these figures.

Consult the National Electrical Code
(NEC) book for a complete list of up-to-
date information. No. 12 copper wire is
probably the most commonly used wire
size. It is often used when a smaller
wire would be approved. Other than
cost, there is no problem with over-
sizing a wire. Under-sizing causes severe problems due to overheating. Fig. 8
illustrates a sample of various wire sizes
for both solid and stranded wires.

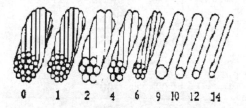

Figure 8. Wire size and type determines
current carrying ability (ampacity).

WHAT IS AN INSULATOR?
Insulators are materials that offer high
resistance to electron flow. Materials
that have 5 or more electrons in the outer
orbit are considered insulators. There is

no perfect insulator. Insulators can break down due to moisture, heat, excess current flow, vibration, chemicals, etc. Insulation can be heat resistant, moisture resistant, oil resistant, etc. The type of insulation or covering determines where the conductor can be safely used. Always use care to avoid damaging the insulated covering on the wire.

The manufacturers of electrical wires use letter codes to designate the type of insulation on a wire. Insulation types are coded by letters of the alphabet and stamped on the insulation surface. The most commonly used types are TW, THHN, THN, or THWN. Consult the NEC handbook for a complete list of insulation types. Fig. 9 shows how conductor insulation is marked.

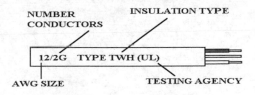

Fig. 9. Insulation information.

WHAT IS A SEMI-CONDUCTOR?

Since a material that has 3 or fewer free electrons is considered a conductor, and a material with 5 or more bound electrons is an insulator, what about a material that has exactly 4 electrons in the valance ring? Silicon is one such material. Pure silicon is not a good conductor because it has 4 electrons in its outer orbit that bond with other silicon atoms to form a very stable crystal.

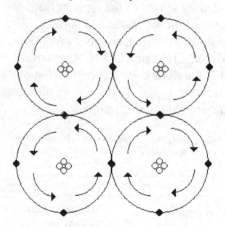

Fig. 10. Silicon atoms

The outer ring of a silicon atom has 4 electrons, but there is room for 8. Therefore, the electrons share orbits with other atoms to form a "covalent" bond. However, if an impurity with either 3 or 5 electrons in the outer orbit is added to the silicon, the crystalline structure will have either an excess electron or a "hole", and can become a conductor or insulator if the correct voltage and polarity are applied. The new structure is called a P-Type or N-Type material.

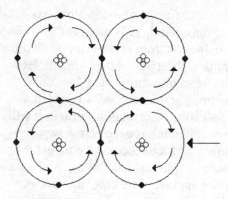

N - TYPE MATERIAL (MISSING 1 ELECTRON)

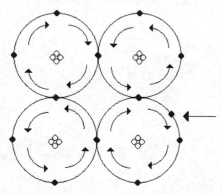

P - TYPE MATERIAL (1 EXTRA ELECTRON)

Fig. 11. P-type and N-type materials.

By sandwiching a piece of N-Type and P-Type material together, an "electrical check valve" can be produced. Electrons would be allowed to flow into the N-type material and out of the P-Type material. However, electrons attempting to enter the P-Type material would be blocked and no current would flow. This simple solid state device is called a *diode.* A diode is represented in electrical schematics by the symbol:

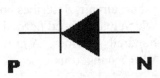

P　　　　**N**

Fig. 12. Schematic symbol of a diode.

HOW DO I PROTECT A CIRCUIT?

Overcurrent occurs when excess current is flowing through a wire. Overcurrent causes the wire to become hot and presents a serious hazard. Overcurrent can be caused by a variety of electrical problems such as; loose connections, ground fault (short circuit), defective resistance, too many loads, etc. An overload describes an overcurrent between two and ten times the normal current. A ground fault (or short circuit) describes an overcurrent which may exceed normal current by hundreds of times. A ground fault is VERY dangerous.

Fuses and circuit breakers are used to protect a circuit against overcurrent. The amperage rating of a fuse must not be greater than the ampacity of the wires being protected. Fuses and circuit breakers are used to detect excessive load current and open the circuit before danger arises. It is standard practice to locate fuses in the main power supply and in each branch circuit. A blown fuse in a branch circuit helps confine the problem to a specific area. Fuses, overloads, and circuit breakers are used to protect wires and equipment, not people.

The amperage and voltage rating of a cartridge fuse determines the physical size of the fuse. The fuse holding devices are sized according to the same procedure. This helps prevent over sizing fuses on circuits designed for a certain maximum amperage. Fig. 13 illustrates fuse dimensions (inches) according to voltage and amperage ratings of the fuses.

AMPERAGE RANGE	250 V FUSE LENGTH (INCHES)	600 V FUSE LENGTH (INCHES)
1/10 to 30	2	5
35 to 60	3	5 1/2
70 to 100	5 7/8	7 7/8
110 to 200	7 1/8	9 5/8
225 to 400	8 5/8	11 5/8
450 to 600	10 3/8	13 3/8

Fig. 13. Amperage and voltage determines fuse size.

Cartridge fuses are available as ordinary fuses (one time blow) or dual-element type. Time delay (dual element) fuses are designed to permit an overload of short duration, but blow instantly if a short circuit occurs. Time delay fuses are necessary when fusing circuits for electric motors. Round type cartridge fuses are used up to 60 amperes. Knife blade contacts are used for fuses over 60 amperes. It is important that the fuse ends make good (tight) contact in the fuse holder. Loose connections or high air temperature around a fuse reduces the amperage rating and causes needless shutdowns.

WHAT ARE LOADS AND SWITCHES?

Manufacturers of electrical devices install the correct resistance for the device to perform the correct amount of energy conversion. It is important to connect these loads to the designed voltage. When connecting a load to a voltage source, a minimum of two conductors must be used. A potential difference (voltage) must exist between the two wires. This power source is connected to each end of the resistance. The potential difference causes electrons to flow through the resistance. Electrons flowing through the resistance causes electrical energy to be converted to another form of energy.

A load cannot operate unless the circuit provides a complete pathway for electrons to flow into and out of the load. Switches are always connected in series with a load (one after the other). More than one switch is often used to control and / or provide safety protection.

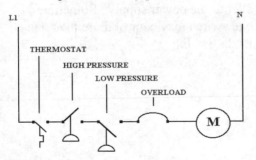

Fig. 14. Multiple switches / safeties.

To provide proper voltage supply, energy conversion devices, are usually connected in parallel. A parallel circuit is connected "from one side of the power supply to the other". The ultimate test of a parallel connection is that the device can be removed without effecting the operation of other devices. In parallel connections, each device is connected independently from all others.

When more than one load is connected to a power source, switches are required to control the individual loads. The switches are connected in series with the loads and each load is connected in parallel to the power supply. Sometimes one switch may control more than one load. See Fig. 15.

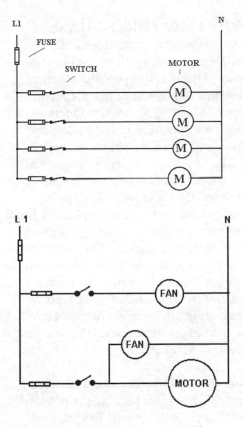

Fig. 15. Switches are connected in
series with a load. Loads are connected
in parallel with each other.

WHAT IS "THREE PHASE"?

The power plant generator rotates three
different conducting loops at the same
time. This is called polyphase generation and is much like three different
power plants. The three conducting
loops are spaced exactly 120 degrees
apart and are called phases or legs.
While one phase (or leg) is magnetically
"positive", another phase is "negative",
and the other is at zero. The three
phases take turns changing polarity from
positive to negative to zero, at a rate of
7200 times per minute. This 3-way posi-
tioning (positive, negative, zero) is continuous as each loop rotates inside the
generator. This method of polyphase
generation produces alternating current
in each of the three phases that are "out
of step" with each other.

A potential difference exists between
any two hot wires because the polarity is
different. Each hot wire has the same
voltage, but different polarity (+ vs. -).
Electrons flow according to polarity
(negative to positive). Therefore, the

potential between any two hot wires is additive because 120 volts (negative) plus 120 volts (positive) equals 240 volts. One wire is "pushing" another is "pulling" with equal force. This is called alternating current (ac), because the hot wires are alternating (taking turns) between positive and negative at a rate of 7200 times per minute. Therefore, the electrons are simply alternating back and forth along with the changing polarity. The voltage tester reads the average voltage of the two wires, called the root mean square (rms). See Fig. 16.

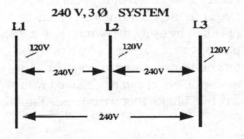

Fig. 16. Voltage between any two wires is additive because polarity is different.

Many electrical loads are designed for connection to all three hot wires. These are called three phase loads. The normal color code for these "hot" wires is black or red. However, they can be any color EXCEPT white or green. The letter "L" (for Line) is used to help identify the three phases (L1, L2, L3).

WHAT IS A SINGLE PHASE SYSTEM?

Some loads are designed to operate with just two hot wires from a three phase system. These are called single phase loads, not two phase. (The term "two phase" refers to an old system that is no longer in use.) The higher voltage is obtained by using two wires from a three phase system. These two wires will alternate from positive to negative. This push-pull effect can be obtained with any two phases (hot wires). See Fig. 17.

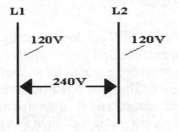

Fig. 17. 240 volt, 1 Ø system.

WHAT IS THE NEUTRAL (WHITE) WIRE?

The Earth is an excellent conductor of electricity, and damp soil is a better conductor than dry soil. The Earth is always at zero potential (no voltage) and can be used to complete an electrical circuit. Many electrical devices operate with just one hot wire from a three phase system and another wire called the neutral. This method is also called single phase, and involves lower voltage. A potential difference exists because the hot wire has voltage and polarity, but the "neutral" wire is connected to the Earth (grounded) which is zero voltage.

While the "hot" conductor usually has black insulation, it can be another color (except white or green) for ease of identification. The neutral wire has white or gray insulation and is connected to a solid copper rod driven eight feet into the ground. This copper rod is called a "grounding electrode". This grounded neutral wire provides a pathway for electrons traveling to and from the Earth, and has zero voltage. The neutral wire is a current carrying conductor, but has no voltage. A potential difference exists between the Neutral wire (no voltage) and the black wire having 120 volts. See Fig. 18.

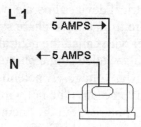

Fig. 18. Amperage flowing in the hot wire also flows in the neutral wire.

WHY DO I NEED A SAFETY GROUND (GREEN) WIRE?

The equipment grounding conductor is another wire that is added for safety purposes, and is called the safety ground. The safety ground is required by the National Electric Code on all newer electrical systems. The color code for this wire is green, or bare copper (not insulated). The safety ground is NEVER connected into the normal electrical circuit. This conductor serves strictly as a "safety valve" when a ground fault occurs in the electrical equipment. The safety ground wire is connected to the frame of a motor or appliance and provides an alternate pathway for electrons to travel to ground and not through someone's body. Many commercial and industrial applications require all electrical wires to be installed inside metal conduit pipes (not plastic). This metal-to-metal pathway should be grounded by connection to a grounding electrode or to the steel framework of the building (which is grounded).

WHAT ARE TRANSFORMERS?

A chief advantage of alternating current is the fact that it can be generated at one voltage, transmitted at a higher voltage, and then reduced to a lower voltage at the point of use. Transformers make it possible to increase the voltage (step up) or decrease the voltage (step down). A transformer has two windings: a primary (incoming voltage) and a secondary (outgoing voltage). Voltage at the secondary is determined by the number of coils or wraps in the secondary winding, versus the number of coils in the primary winding. The transformer primary terminals are normally labeled H1, H2, H3, and the secondary terminals are tagged X1, X2, X3, (X0 is used to indicate the neutral terminal).

Single phase transformers are rated by VA (volts x amps) at the secondary. Transformers rated over 1000 VA are normally rated KVA, with K representing 1000. An overloaded or undersized transformer will burn out because the secondary coil cannot carry the required current. The secondary winding is considered to be a power source for any loads connected to the transformer. Single phase transformers normally have two wires for the primary (incoming voltage) and two wires for the secondary (outgoing voltage).

Three phase transformers can be of the Wye or Delta type, although the Wye (or Star) is most popular. Because electricity is produced in three phases, a variety of Wye, Delta, Wye-Delta, or Delta-Wye transformers are used. This provides a variety of voltages that are in common use. See Fig. 19 A, B, C, and D.

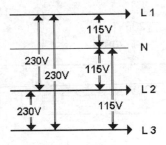

Fig. 19 (A). 230V, 3Ø, 4-wire.

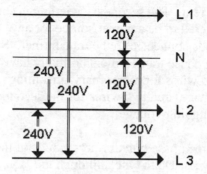

Fig. 19 (B). 240V, 3Ø, 4-wire.

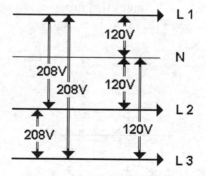

Fig. 19 (C). 208V, 3Ø, 4-wire.

44

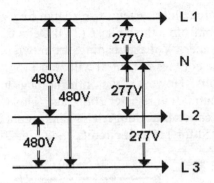

Fig. 19 (D). 480V, 3Ø, 4-wire.

Some older systems use a High Leg system from a three phase (3 Ø) Delta transformer. Voltage readings from two of the hot legs to neutral will read 115 volts. However, one of the hot legs to neutral will register 208 volts. This wire is called the "Stinger" leg and cannot be used for 115 volt circuits. See Fig. 20.

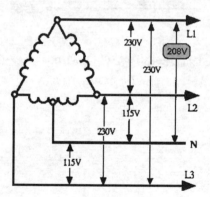

Fig. 20. Delta High Leg System.

Another older system is the Dead Leg system 3 Ø Delta transformer. One corner of the Delta transformer is "grounded" and called the Dead Leg. A voltage reading of 240 volts can be obtained between any two phases (including the Dead Leg). A voltage reading from the Dead Leg to ground is zero volts. See Fig. 21.

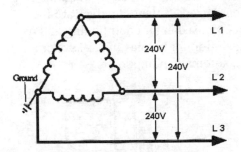

Fig. 21. Delta Dead Leg System.

WHAT IS A SOLENOID VALVE?

Electrically operated valves are called solenoid valves. When current flows through the coil, electromagnetism is produced and lifts the valve's plunger. This opens the valve and liquid flows through the valve. When the coil is de-energized, magnetism disappears and the valve closes (normally closed). Some solenoid valves are designed to be normally open, and will close when the coil is energized. Solenoid valves are frequently used in air conditioning and refrigeration for hot gas bypass systems and automatic pumpdown.

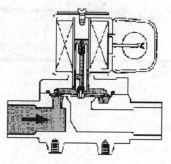

Fig. 22 Solenoid valve

WHAT IS A RELAY?
A relay uses a magnet to operate a switch (or contacts). Relays are often used to control electrical loads from a remote location. When the relay coil is energized, electromagnetism causes the contact (switch) to open (or close). The electrical circuit to the relay coil is entirely separate from the circuit through the relay contact. The coil voltage (control circuit voltage) may be 24 volts and the circuit through the relay contacts may be 120 volts. Thus, low voltage is used to control a switch that controls a high voltage load. See Fig. 23.

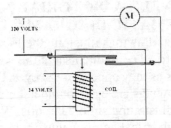

Fig. 23. The coil voltage is separate from the circuit to the motor.

Relays often contain more than one set of contacts or switches. Relay contacts are always shown as normally open (N.O.) or normally closed (N.C.), with the coil in the de-energized position. Energizing the coil causes the contacts to change position. The relay contacts have low current ratings, with a maximum of 10 amps being considered normal. Current flow in the coil circuit is very low (often less than 1/4 ampere). A 24 volt control circuit is safer to personnel, permits the use of smaller wire, and causes less arcing at the switch.

WHAT IS THE DIFFERENCE BETWEEN A PICTORIAL AND SCHEMATIC DIAGRAM?

Pictorial drawings illustrate the location of electrical components and the wiring circuits as they appear to the eye. Pictorial drawings are easy to understand when illustrating simple circuits. When many electrical components are involved, the pictorial diagram becomes too difficult. The schematic diagram presents an orderly method of illustrating electrical circuits. The schematic does not illustrate where the components are located, but only how they are connected in the circuit. A schematic drawing is less cluttered and uses symbols to illustrate components. Experienced technicians prefer the schematic method because the circuits are easily traced and troubleshooting is easier. To use a schematic diagram, the technician must know the equipment components and their corresponding electrical symbols. Most schematic diagrams contain a legend that describes any abbreviations and special components. Fig. 24 illustrates the pictorial and schematic drawing of the same circuits (wire for wire).

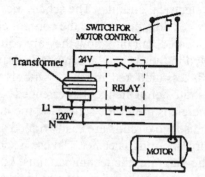

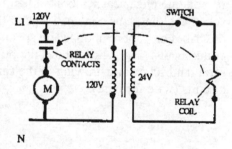

Fig. 24. Pictorial and schematic view of using a relay for motor control.

WHAT IS A LADDER DIAGRAM?

Another method of drawing a schematic is the ladder diagram. A ladder diagram is arranged with the power supply lines drawn vertical as the legs of a ladder. Each horizontal line or rung of the ladder contains one load and its control switches. Each load line may be numbered for ease of identification. Control switches are usually located on the line side of the device being controlled, however some manufacturers will choose to place a switch or two on the load side of the rung.

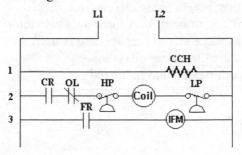

Fig. 25. Ladder Diagram

WHAT IS A CONTACTOR?

A contactor is an electrical switching device and operates much like a relay. However, the contactor has heavy duty contacts for controlling larger loads. The heavy duty contacts are normally open (N.O.) and controlled by a magnetic coil. When the coil is energized, the heavy duty contacts close. The contacts are used to open or close a circuit between the main power supply and the load. Contactors are rated according to the maximum amperage flow through the contacts for a specific voltage. A 2-pole contactor has two separate contacts and is used for controlling 240 volts, single phase circuits (residential air conditioning). A three pole contactor has three sets of contacts and is used for controlling three phase loads (commercial and industrial).

Fig. 26. 3-pole contactor (Honeywell).

It is normal for contacts to become pitted
and burned due to arcing when contactors open and close. Use of a file or
sandpaper to clean the contacts is not
recommended. Such "cleaning" de-
stroys the contact surfaces and increases
arcing. Replacement contacts are usually available from a local supplier.

The numbering system for contactors determines direction of the current flow through the contacts. It is standard policy for power supply to enter at the top of the contacts and the load is connected to the bottom of the contacts. The line power (inlet) terminals are labeled L1, L2, L3, and the load (outlet) terminals are labeled T1, T2, T3.

WHAT IF THE CONTACTOR COIL BURNS OUT?

The magnetic coil is located inside the contactor, but is NOT electrically connected to the main contacts. Because the coil is a separate device, the coil voltage can be different than voltage at the main contacts. The coil has its own terminals for making electrical connections and replacement of the coil is easy. Because the coil is electrically separate from the main contacts, it is common practice to use lower voltage to operate the coil. Residential systems generally use 24vac. Lower voltage and current in the control circuit is safer and permits the use of standard switching devices. A step-down transformer is often used to obtain the lower control circuit voltage.

Any number of contacts and safety control switches can be located in the control circuit. The contacts for these controls are connected in series with the coil and may include overloads, thermostats, pressure controls, fuses, limit switches, flow controls, or oil pressure controls. All switches in the control circuit must be closed before the coil can be energized. However, any switch can disconnect power to the coil and stop the motor. Many safety controls are manual reset, while others are automatic reset.

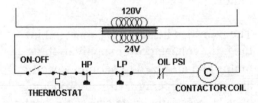

Fig. 27. All the control switches must be closed to energize the coil. Any switch can open and stop the motor.

WHAT IS A LINE STARTER?

A line starter is basically a contactor with built-in overload protectors. A line starter is often used to operate and protect three phase motors. The overloads serve to protect the motor against excess amperage (overload) and are more accurate than fuses. The overload protectors are normally connected to the bottom of the contactor and installed prewired at the factory. One overload is required for each phase of power supply to the motor (L1, L2, L3). Supply must flow through the heavy duty contacts and then through the overload "heater" before traveling to the motor.

The overload heaters are connected in the high voltage power supply to the motor and are sized to permit a specific amount of amperage without producing heat. Excess amperage causes the heater to generate heat that will cause a nearby set of contacts to open. The overload contacts are connected in the control circuit supplying power to the contactor coil. Fig. 28 shows a cutaway view of one leg of the contactor circuit and the connecting overload.

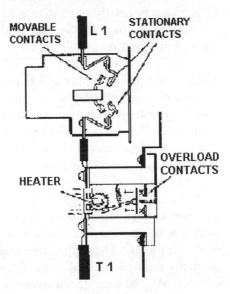

Fig. 28. Cutaway view of a contactor and overload.

A manual reset is provided to re-close the overload contacts after the heater cools off. Heaters are available in a wide variety of amperage capacities. Heaters are normally sized to provide accurate motor protection at slightly above Full Load Amps (FLA).

Three phase power supply to the motor is connected through the heavy duty contacts, through the heater, and then to the motor. The overload switches are connected in the control circuit to the contactor coil. Excess amperage in a high voltage leg causes the overload contacts to open. Opening of the overload contacts diconnects the power supply to the contactor coil and thus stops the motor. The tripped overload normally requires manual reset of the contact. Any of the three heaters can open the one contact.

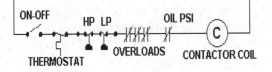

Fig.29. Overload circuit protection

WHAT ARE INDUCTION MOTORS ?

Over 90% of all motors are the induction type and operate on alternating current. DC motors and their controls are not discussed in this booklet. The induction motor is most widely used and its operation is fully explained. Knowledge of the induction motor is easily applied to other motor types. Sometimes a motor malfunction serves as a warning of other system problems. A good working knowledge of motors and their operation is necessary to properly troubleshoot an AC&R system and perform required repairs.

WHAT ARE THE PARTS OF A MOTOR?

There are two main parts of any motor: the rotor (part that rotates), and the stator (stationary electromagnetic coils arranged in a circular pattern). The rotor is placed inside the circular electromagnets. Endbells (with bearings) are used on each end of the motor and the entire assembly is bolted together.

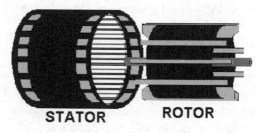

STATOR   **ROTOR**

Fig. 30. The rotor and stator are the two main parts of a motor.

Induction motors operate on the principle of induced magnetism. The rotor is placed in the center of the stationary electromagnets (stator). When current flows in the stator coils, a strong magnetic field is produced in the stator poles. This stator magnetism induces opposite magnetism in the rotor.

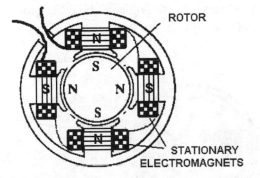

Fig. 31. The rotor magnetism is attracted (and repelled) by the rotating stator pole.

WHAT ARE STATOR POLES?

Two (or more) stationary electromagnets (called poles) are positioned at opposite sides of a circle inside the motor. A strong magnetic field is produced when current flows through the coils. In a two pole motor, the stator poles have opposite polarity because the coils are wrapped in opposite directions. One coil produces a North pole and the other produces a South pole.

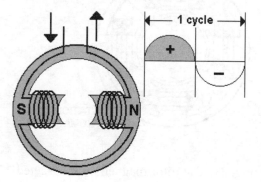

Fig. 32. Coils are wrapped in opposite directions to produce opposite polarity in stator poles.

An electromagnet has two distinct advantages: 1. Its core is magnetized only when current flows through the coil, and 2. The polarity of an electromagnet can be changed by reversing current flow through the coil.

Alternating current automatically reverses polarity of the stator poles at a rate of 120 times per second (one positive and one negative per cycle). When alternating current reverses, the polarity of each stator changes. Thus, polarity of the stator electromagnets automatically change due to alternating current flow.

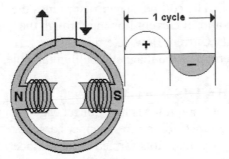

Fig. 33. Polarity changes 120 times/sec.

WHAT IS A STATOR?

Motors have stationary coils of copper wire (main windings) which are carefully wrapped around layers of soft iron (called poles). These magnetic poles (coils and laminated cores) are permanently mounted inside of the motor shell. A minimum of two poles (N and S) are required. Each pole is located exactly 180 degrees around the circle. This arrangement of magnetic poles is called the stator.

The size (AWG) of copper wire used and the number of wraps in the coil determines the amount of resistance in the coil. The coil resistance determines the amount of current flowing through the coil. Current flow determines the strength of the magnetic field in the pole.

WHAT IS A ROTOR?

A common rotor is the squirrel cage type. Instead of wires, copper bars are inserted into slots formed in the surface of the core. The ends of these copper bars are joined together, thus forming a series of closed loops arranged in a sort of squirrel cage, hence its name.

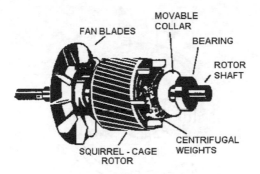

Fig. 34. Typical squirrel cage rotor for single phase motor.

The fields created by the stator electromagnets cut across the closed loops in the rotor and large currents are induced in the rotor loops. These induced currents create a magnetic field in the rotor that is opposite polarity of the stator electromagnet. Because opposite magnetic poles attract, the rotor is locked into a fixed position. If the rotor is given a spin, it will continue to spin due to attraction and repulsion of the alternating polarity of the stator poles.

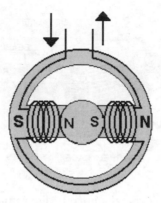

Fig. 35. Rotor polarity is opposite of stator poles.

Once started, the rotor continues to spin because the rotor's South pole is attracted by the stator's North pole. Also, the rotor's South pole is repelled by the stator's South pole. This push-pull action is continuous as the poles reverse polarity and the rotor tries to catch up with the changing polarity.

WHAT ARE SPLIT PHASE MOTORS?

Single phase motors cannot start with a single stator winding, called the run (or main) winding. The delay in rotor start-up permits the stator poles to return to original polarity. This attraction of unlike poles results in a condition called locked rotor (cannot turn). If the rotor shaft is given a spin manually, the motor runs properly. The rotor will turn in the direction spun. However, when the motor stops it cannot re-start unless given another manual spin.

A start winding is required to have automatic starting. The start winding establishes another magnetic field in the stator that is "out of step" with the main winding. These are called split phase motors.

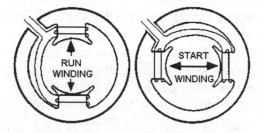

Fig. 36. The split phase motor requires a start winding to automatically start the rotor turning.

The start winding is made of smaller diameter wire than the run winding and has more turns (wraps) on the laminated pole. This higher resistance in the start winding produces a magnetic field that is slightly behind the main winding.

At start-up, current flows through both windings. Magnetism in the start winding is slightly behind the run winding due to its higher resistance. These two magnetic fields, one behind the other, create the necessary turning force on the rotor.

The magnetic field produced by the start winding is strong when the main winding is charged from N to S. This produces an "on" and "off" condition between the start and run windings. When one is "on" the other is "off". This provides the necessary torque on the rotor to make it turn.

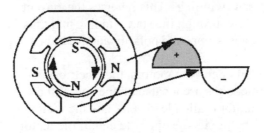

Fig. 37. The start winding creates a magnetic field that is out of step with the run winding.

HOW DO I CHANGE THE DIRECTION THAT THE MOTOR ROTATES?

Rotation (clockwise or counter clockwise) is always determined by the direction of current flowing through the start winding. To reverse rotation, reverse the two power supply connections to the start winding. This reverses the current flow through the start winding and thus causes opposite polarity and rotation.

On open type motors, the electrical connections are located at one end of the motor (called lead end), and the motor shaft exits the opposite end of the motor (called shaft end). Direction of rotation is normally determined by viewing the shaft end, with the lead end the farthest away. However, rotation for General Electric motors is determined by viewing the lead end and having the shaft end farthest away.

Direction of rotation should always include which end of the motor is being viewed. Rotation of the shaft is com-

pared to the movement of a clock. If the
shaft is turning to the right, rotation is
clockwise (CW). If the shaft is turning
to the left, rotation is counter clockwise
(CCW).

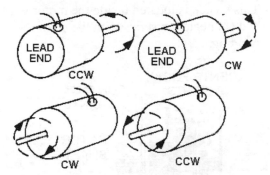

Fig. 38. Rotation of motor shaft depends
upon which end of the motor is being
viewed.

WHY MUST THE START WINDING BE DISCONNECTED?

The only purpose of the start winding is to get the motor started. The start winding will burnout if permitted to remain in the circuit. Single phase open type motors use a centrifugal switch for disconnecting the start winding after start-up. This switch is located inside the motor and connected in series with the start winding.

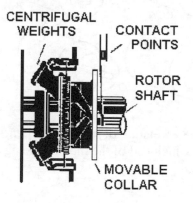

Fig. 39. A centrifugal switch if often used for disconnecting the start winding after start-up.

The contacts on the centrifugal switch are closed when the motor is not running. When the motor achieves 75 percent of speed, enough centrifugal force is produced to a pair of weights to swing outward and open the switch. When the motor stops, springs pull the weights back and the switch is closed for the next start-up. Failure of the centrifugal switch to open permits the start winding to remain in the circuit. This results in high amperage and the motor overload should stop the motor.

WHAT DETERMINES MOTOR SPEED?

The speed of a motor is determined by the number of stator poles. The rotor speed is rated in revolutions per minute (RPM). Synchronous speed is figured by dividing the number of stator poles into 7200 (alternations per minute). In a motor running at full load, the rotor actually rotates at a speed about 4 to 5 percent below synchronous. The difference in motor speed is called slip.

NUMBER OF POLES	SYNCHRONOUS SPEED	ACTUAL SPEED
2 - POLE	7200 ÷ 2 = 3600 RPM	3450 RPM
4 - POLE	7200 ÷ 4 = 1800 RPM	1725 RPM
6 - POLE	7200 ÷ 6 = 1200 RPM	1150 RPM

Fig 40. Motor speed table.

HOW DO I CALCULATE MOTOR HORSEPOWER?

Motors are rated by the amount of torque they can produce. This turning power is rated in horsepower. The United States system rates horsepower according to the power consumption used by the motor. This power is rated in Watts, and 746 Watts equals one hp. Thus, a five horsepower motor will use 3730 watts (5 X 746=3730). This formula assumes 100% efficiency. When the wattage and voltage are known, Ohm's Law makes it possible to determine the amount of amperage draw at full load conditions.

Motors of less than one horsepower are called fractional horsepower motors. For example, 1/2 hp and 1/4 hp are fractional horsepower motors.

WHAT ARE LOCKED ROTOR AMPS? (LRA)

At start-up, current flowing in the motor is determined by the resistance of the windings. Starting current is about six times higher than normal running amperage. This high current flow is called locked rotor amps (LRA). As the motor picks up speed, counter-EMF is generated and reduces current flow. At full speed, current flow is determined by the resistance of the run winding and the counter-EMF generated by the motor.

WHAT ARE FULL LOAD AMPS? (FLA)

Full load amps (FLA) refers to the amperage the motor draws when it is at normal speed and fully loaded. Most induction motors operate at less than FLA because the motor is rarely working at fully loaded conditions. An overload occurs when amperage exceeds the FLA rating on the motor data plate.

WHAT IS AN OVERLOAD PROTECTOR?

An overload protector is a device that protects a motor against overload conditions. A variety of overload protectors are commonly used. The type used depends greatly upon the type of motor and its application. A common motor overload for fractional horsepower, single phase, AC motors uses a snap-acting bimetallic disc to make and break a set of contacts.

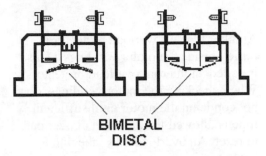

**BIMETAL
DISC**

Fig. 41. Overload protectors normally
use a snap-acting bimetal disc and break
a set of contacts.

Excessive current causes the bimetal
disk to deflect and open the circuit .
When cooled the disc returns to the
closed position.

Repeated opening and closing of the
overload is called cycling on the overload. Tripping of the overload protector
is a warning that the motor is in danger
of a burnout and it is important to isolate
and repair the cause of the overload.

Another type of overload protector is buried inside the motor windings. This is called an internal overload protector. When checking the motor windings, do not condemn the motor until sufficient time is allowed for the internal overload to reset. An overload may take 4 to 8 hours to reset.

WHAT ARE CAPACITORS?

A capacitor is often used to improve the operating characteristics of single phase motors. A capacitor is an electrical device that stores electrical energy. Capacitors are used to increase and decrease the magnetic field produced by the motor start winding. These two uses result in two different names for single phase motor capacitors: start capacitors and run capacitors.

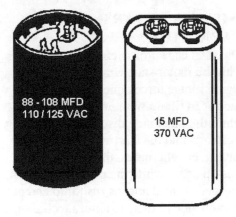

Fig. 42. Start and Run capacitors.

Many capacitors have a bleed resistor
soldered or connected to the terminals.
The resistor will safely discharge the
capacitor each time the circuit opens so
it will not overcharge the next time voltage is applied. The resistor also reduces
the severity of arcing that occurs at the
start switch contacts. Since the charge
has been bled from the capacitor, the
possibility of a technician being shocked
when it is removed from the circuit is
greatly reduced.

WHAT IS A START CAPACITOR?

The start capacitor is connected in series with the motor start winding. The applied voltage forces one side of the capacitor to fill up with excess electrons while the other side discharges its excess electrons into the start winding. When the current alternates, the empty side fills up again while the other side discharges. The electrons rush into and out of each side of the capacitor, according to the alternations of ac current. When one side is full, the other side is empty.

One side of the start capacitor discharges excess electrons into the start winding. This increases current flow in the start winding and increases the strength of the magnetic field. This provides better starting torque. The start capacitor is not designed to stay in circuit after the motor starts. The start capacitor is typically energized for between 0.75 and 1 second and should never exceed 4 seconds. Start capacitors are easily damaged and are designed to tolerate about 20 starts per hour without overheating. Start capacitors are usually constructed with a plastic "bakelite" case.

Replacement capacitors should have the same microfarad and voltage rating as the one being replaced. In an emergency situation, a replacement start capacitor up to 20% over capacity may be used as a temporary repair until the proper capacitor can be installed. Never install a capacitor that is under capacity.

WHAT IS A RUN CAPACITOR?

Run capacitors are better built and almost trouble free. The capacitance rating is low and it is built to remain in the circuit constantly. The case is made of heavy gauge metal, which helps dissipate heat. The run capacitor increases motor running torque by keeping the start winding slightly energized during the run cycle. The run capacitor is connected in series with the start winding, just like a start capacitor. However, the run capacitor is not disconnected. The run capacitor limits the amount of electrons entering the start winding.

Many run capacitors have an identifying mark on one of the terminals. The mark shows which plate is closest to the metal case. The hot wire should be wired to this terminal. In the event that the capacitor becomes shorted to the case, the fuse or circuit breaker will open. This will prevent the motor from operating without both of its windings, limiting the potential for overheating and damage.

HOW ARE CAPACITORS RATED?

Both capacitors have two important ratings. One rating is VAC (volts, alternating current), and the other is microfarads (MFD or μf). Microfarads indicate the capacitor's energy storage capacity. Start capacitors have much higher microfarad ratings than run capacitors. Run capacitors are available from 1.5 to 70 MFD, while start capacitors range from 21 to 1,600 MFD.

HOW CAN I TEST A CAPACITOR?

Capacitors can be tested using an ohm-meter or a specially designed capacitor tester. When testing a capacitor with an ohmmeter, the capacitor must first be discharged. In order to prevent damage to the capacitor, do not short across the terminals with a screwdriver. The sudden surge of current through the direct short circuit can cause unnecessary stress to the plates and dielectric in the capacitor, or it may cause the small internal wires that are soldered between the plates and terminals to burn open. Damaging the capacitor during testing is a sure way to lead to a misdiagnosis.

Instead of a screwdriver, use a 15,000 ohm resistor to jumper across the capacitor terminals. This will allow the current to dissipate slowly and avoid damaging the capacitor.

Some capacitors are fitted with a bleed resistor. The resistor bleeds off the charge each time the circuit is de-energized. The bleed resistor must be carefully removed before continuing with testing.

Once the capacitor has been safely discharged, connect the leads of an ohmmeter to the terminals. The meter needle should deflect slightly towards 0 ohms, then return to an infinite reading. Since an ohmmeter has its own power source, the meter will charge the capacitor slightly. This will be indicated by a deflection in the meter needle, or a change in the digital readout. When the leads are reversed, the charging action will also reverse and the needle will again deflect toward 0 ohms and return to an infinite reading.

If the needle does not deflect, the capacitor is open. If the needle deflects and remains at 0 ohms, the capacitor is shorted. Connecting the leads between either terminal and the metal case of a run capacitor should show an infinite reading.

This ohmmeter test is valid in diagnosing open or shorted capacitors, these are definitely bad. However, because a capacitor passes the ohmmeter test, you cannot guarantee the capacity rating. To test the capacity rating, a special capacitor tester can be used. If you do not have a capacitor tester, you can measure and calculate the capacitance in microfarads (µf) by using an ammeter and a test cord as shown in Fig 43.

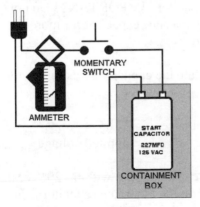

Fig. 43 Testing a start capacitor.

CAUTION: Place the start capacitor in a containment box before energizing. In the event of a failure, a start capacitor can overheat and expel flaming material. Before connecting the test cord, perform the ohmmeter test. If the capacitor is shorted, do not proceed with this test!

Close the momentary switch for 3 seconds and record the current flowing into the capacitor. IMPORTANT! Do not energize a start capacitor for more than 3 seconds.

Calculate the capacity by using the formula:

Capacitance μf = $\dfrac{2.652 \text{x Amps}}{\text{applied voltage}}$

If the capacitance is within +20% of the rating on the label, the capacitor is in good condition and can be used to start the motor.

If the capacitance is below the label rating, the capacitor is breaking down and must be replaced.

Run capacitors are tested using the same procedures as described for a start capacitor, with a few exceptions. A run capacitor is housed in a metal case. Therefore, when ohmmeter testing, check that there is no continuity from either terminal to the case. Because a run capacitor is designed for continuous duty, there is no need for a momentary switch to be wired into the test cord as with the start capacitor.

WHAT IS A CSIR MOTOR?

Single phase motors having a start capacitor are properly called a Capacitor Start-Induction Run (CSIR) motor. The CSIR motor is used when the motor must start under loaded conditions. Extra torque is required for starting purposes only. Once started, the capacitor and the start winding must be disconnected from the circuit.

WHAT IS A PERMANENT SPLIT CAPACITOR MOTOR?

The permanent split capacitor (PSC) motor is often used to operate fans and blowers. At full speed the fan blades are fully loaded due to the weight and volume of air being moved. The PSC motor uses a "run" capacitor to keep the motor start winding slightly energized. Thus, the start winding is used to assist the run winding during fully loaded conditions.

Run capacitors have low microfarad ratings and act as a throttling device to limit the amount of electrons flowing into the start winding. The run capacitor is connected into the start circuit permanently, hence the name Permanent Split Capacitor.

WHAT ARE MULTI-SPEED MOTORS?

Single phase, multi-speed motors are very common. Motor speed is determined by the number of stator poles being connected to the power source. The multi-speed motor has several external taps (wires) for selection of motor speed desired. High speed uses two stator poles (3600 RPM), medium speed uses four stator poles (1800 RPM), and low speed uses all six stator poles (1200 RPM).

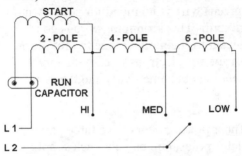

Fig. 44. The multi-speed motor has several wires for selecting desired speed.

WHAT ARE THREE PHASE MOTORS?

Three phase motors are very common in commercial and industrial applications because they are smaller and more efficient than single phase motors of equal horsepower. Three phase motors have high starting torque and high running torque without the use of a start winding or capacitors.

For proper operation, all three supply wires (L1, L2, and L3) are connected to the motor terminals. The safety ground (green) wire is included for equipment ground. This grounding wire is connected to the motor frame to provide an escape for electrons in case the motor windings become shorted to the metal frame.

Three phase motors have three pairs of stator poles, one pair for each supply wire. Each winding produces a North and South pole, which is called one pole per phase. A typical three phase motor has three pairs of stator poles, thus having three Norths and three Souths. Each North and South combination is located directly opposite each other. These stationary poles are equally spaced in a circle, exactly 60 degrees apart.

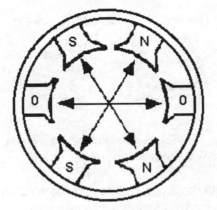

Fig. 45. Pictorial view of three phase
motor having three pairs of stator poles,
one pair for each supply wire.

The resistance of all three windings in a
three phase motor are the same. Only
one end of each winding is brought outside the motor for connection to the
power source. The other end of each
motor winding is factory connected inside the motor. See Fig. 46.

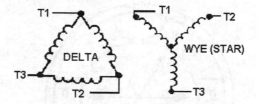

Fig. 46. Schematic symbols of a Delta and Wye connected 3-phase motor.

With three phase alternating current, the three power supply wires take turns changing their polarity from North to South to zero. When one wire is North, another wire is South, and the third wire is zero. This changing of polarity in the supply wires produces a strong rotating magnetic field in the stator poles that are out of step with each other. The alternating zero does not produce polarity in the stator poles, which permits the other two poles to produce the rotating push-pull effect on the rotor. See Fig. 47.

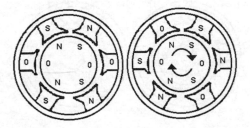

Fig. 47. Stator polarity produces a push-pull effect on the rotor.

HOW CAN I CHANGE THE ROTATION OF A THREE PHASE MOTOR?

Direction of rotation is determined by the direction of the rotating field. Reversing rotation on a three phase motor is accomplished by changing any two supply wires. This simple procedure causes the magnetic field to rotate in the opposite direction.

Wrong rotation can be devastating to equipment and personal safety. If necessary, disconnect equipment before checking proper rotation of a motor.

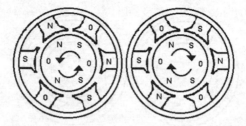

Fig. 48. Changing any two supply wires will change rotation of rotor.

HOW DO I CHECK RESISTANCE OF WINDINGS?

The motor windings on a three phase motor can be checked with an ohmmeter. The motor must be disconnected from the circuit and the readings are obtained from one motor lead to another. If a resistance reading of zero occurs the winding is shorted. If a reading is obtained from any motor lead to ground the winding is grounded. A reading of infinite resistance indicates that the winding is open. In each of these cases, the motor must be re-wound or replaced. The resistance reading on three phase motors windings will vary from less than one ohm to 50 ohms, depending on the size of the motor. The larger the motor, the smaller the resistance. Each winding will have the same resistance, except for the dual voltage type. The dual voltage motor has three "extra" windings and their resistance is one-half the resistance of the main windings.

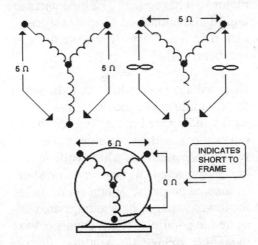

Fig. 49. An ohmmeter is used to check motor windings.

97

WHAT ARE DUAL VOLTAGE THREE PHASE MOTORS?

Many three phase motors are designed for connection to either of two different voltages (240/480). These are called dual voltage motors. Instead of having just three external wires to connect, this motor will have nine. The nine wires are tagged and numbered #1 to #9 for easy identification. NEVER remove these numbers.

Dual voltage motors have an extra set of three windings and the two wires for each winding are brought outside the motor along with the original three wires. Regardless of which voltage is being connected, all nine wires must be connected properly. When connected to the lower voltage, the windings are connected in parallel . When connected to the higher voltage, the windings are connected in series. The three power supply wires are ALWAYS connected to motor numbers T1, T2, and T3.

Fig. 50A illustrates both low and high voltage connections for a Wye (Star) connected, three phase, dual voltage motor. Fig. 50B illustrates both high and low voltage connections for a Delta connected, three phase, dual voltage motor.

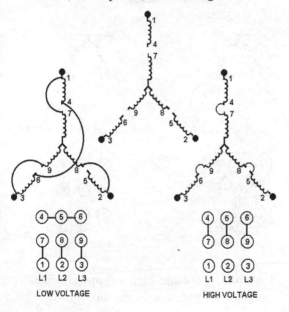

Fig. 50A. Low and high voltage connections for a WYE (Star) connected motor.

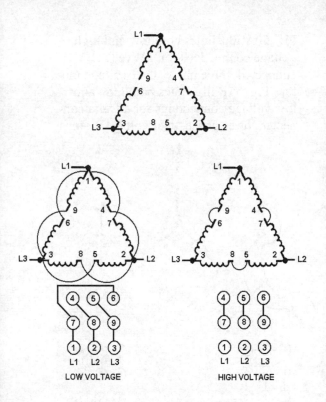

Fig. 50B. Low and high voltage connections for a Delta connected motor.

WHAT INFORMATION IS ON A MOTOR NAME PLATE?

Instructions for making electrical connections to a motor are normally included on the motor nameplate (also called data plate). The nameplate should be carefully viewed before selecting, replacing, or wiring a motor. Fig. 51a. and 51b. show an example of a single phase motor nameplate and a three phase motor nameplate.

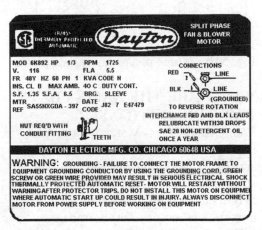

Fig. 51a. Single phase motor data plate.

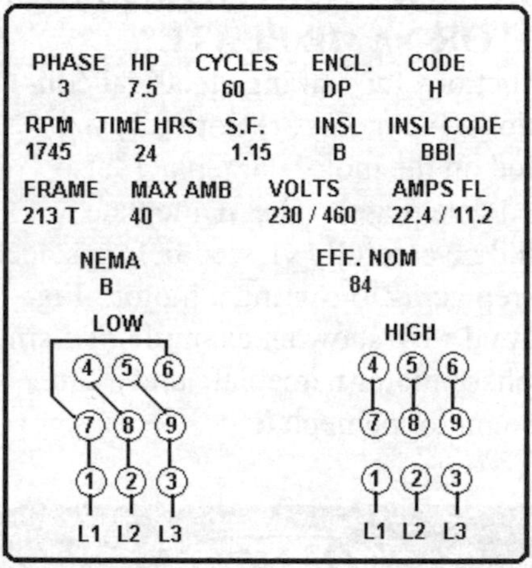

Fig. 51b.. Three phase motor data plate.

WHAT ARE THE NAME PLATE DATA DEFINITIONS?

Frames and Type: Motors of a given horsepower rating are built in a certain size of frame or housing. NEMA has standardized the frame size and shaft heights to be used for each integral horsepower motor. This permits easy replacement or interchanging of motors.

Max Amb: The maximum ambient temperature at which a motor can be operated.

Temperature Rise: The amount of temperature rise permitted above ambient air at rated load.

Duty (Time / Hrs): All electric motors are designed for either Continuous or Limited duty. Those designed for continuous duty will deliver the rated horsepower for an indefinite period without overheating. Limited duty motors are designed to deliver rated horsepower for a specified period of time and will overheat if operation is extended. Limited duty motors are often used to operate valves, pumps, or louvers.

Thermal Protection: Indicates type of thermal protection provided, if any.

FLA (Amps FL): The rated current (in amperes) when operated at full load.

LRA (Locked Rotor Amps): The rated current (in amperes) if rotor is unable to turn.

KVA Code: A letter indicating the starting current required. The higher the locked-rotor-kilovolt-ampere (kVa), the higher the starting current surge.

Insulation Class (INSL): A designation for the type of insulation used. Primarily used for rewinding purposes.

Service Factor (S.F.): The amount of overload that the motor can tolerate on a continuous basis at rated voltage and frequency.

Glossary
A

Alternating current (ac) Electrical
voltage and current that changes
in magnitude and direction in a
cyclic characteristic. The ac
electricity in North America
cycles 60 times per second.

Ammeter An instrument used to measure
current flow through a circuit.

Ampere The unit of current flow equal
to one coulomb of charge moving
through a circuit in one second.

Analog meter A meter that displays
information using a needle
moving over a scale.

Atom The smallest unit of an element,
made up of protons, electrons
and neutrons.

B

BTU Stands for British Thermal Unit. A unit of energy based on the amount of heat required to raise the temperature of one pound of water one degree Fahrenheit.

C

Capacitance A measure of the energy stored in a capacitor. The units of capacitance are farads.

Capacitor A device made with two conductive plates separated by a dielectric. It can be used to boost voltage for a single phase motor.

Capacitor– start motor a modification of a split phase motor that uses a capacitor in series with the start winding. The capacitor causes a phase displacement for starting.

Capacitor-start induction-run motor
 capacitor assists start under high
 load conditions. Once started, the
 motor operates on the run
 winding only.

Capacitor-start capacitor- run motor
 Utilizes a start capacitor for high
 starting torque and a run
 capacitor for running efficiency.

Cathode the negative terminal of an
 electrical device.

Centrifugal switch a switch that uses a
 combination of weights and
 springs that will open its contacts
 at approximately 75% of a motors
 RPM. The switch is wired in
 series with the start winding and/
 or start capacitor.

Circuit electrical wires and components
 which allows current flow in a
 complete loop from the energy
 source back to the energy source.

Circuit breaker a safety device designed to automatically open a circuit at a predetermined overload of current.

Circuit, parallel a circuit arrangement in which all electrical devices have their positive connections connected to one conductor and all negative connections connected to the other conductor. Voltage supplied is the same to each device in the circuit.

Circuit, series an arrangement in which the current must pass through all loads and then back to the energy source. A different voltage will be supplied to each load in the circuit. Current will be the same in any part of the circuit.

Clamp-on ammeter current measuring device that senses the strength of the magnetic field around a conductor.

Coil   wire arranged in a spiral shape (usually around an iron core) that will create a strong magnetic field when current passes through it.

Common   a terminal, connection or other part of an electrical circuit that is shared by other parts of the circuit.

Conduit   a tube used to carry and protect electrical wires. Can be metallic or non-metallic.

Connection   mechanical or electrical joining of two parts.

Contact points   movable points, usually made of tungsten, platinum, or silver, that complete a circuit when pressed together.

Continuity   an unbroken line or continuous path through which electricity may flow.

Control a device that directly operates electrical supply to the equipment. May be automatic or manually operated.

Coulomb the quantity of electricity equal to a current of one ampere in one second. One coulomb equals 6.25×10^{18} electrons passing a point in one second.

Counter electromotive force action which takes place in motors when voltage (EMF) is self- induced in the rotor conductors which is opposite of the source voltage.

Covalent bond atoms joined together to form a stable molecule by sharing electrons.

Current movement of electrons in a conductor. It is usually expressed in amperes. Represented by the symbol "I" for intensity.

Current relay a relay that is operated by the starting current of a motor. It allows the start winding to drop out of the circuit after the motor starts.

Cycle the voltage generated by the rotation of a conductor through a magnetic field, from a zero reference, in a positive direction, back to zero, in a negative direction, and back to zero again. One complete cycle equals 360 degrees of rotation.

Cycles per second the number of alternating current waves in one second of current flow.

D

Data plate equipment identification label, usually containing such items as model, serial number, voltage, and amperage.

DC direct current.

Dead describes a portion of a circuit having no voltage.

Dead leg the grounded phase of a three phase delta wound transformer.

Dead short a very low resistance connection that allows an unrestricted flow of electrons.

Dedicated circuit a circuit that is fused and supplies power to one appliance only.

De-energize to stop electron flow to a device.

Delta transformer a three phase transformer that has the finished end of one winding connected to the finished end of the second. The configuration resembles the Greek letter delta (Δ).

Diac a semiconductor most often used as a voltage-sensitive switching device.

Dielectric an insulating material separating the conducting surfaces of a capacitor.

Digital voltmeter a voltmeter that uses direct numerical display as opposed to a meter movement.

Diode a solid state device that allows current to flow in one direction only. It will rectify alternating current to direct current.

Direct current (DC) current that flows in one direction only in a circuit.

Distribution center an electrical panel that supplies electricity to several places in a structure.

Doping adding an impurity to a semiconductor to produce a desired change in electrical properties.

Double- pole breaker switch a circuit breaker that is used to disconnect both hot wires with a single on-off action.

Double-pole double-throw switch a switch with two poles and two contacts for each pole. Two contacts are always open and the other two are always closed.

Double-pole single-throw switch a switch arranged so that both switches are either open or closed.

Duty cycle the relationship between operating time and off time. Duty cycle of a motor is usually referred to as continuous or intermittent.

E

E   a symbol for voltage (electromotive force)

EER energy efficiency ratio

ELI a term used to remember that voltage (E) in an inductive (L) circuit leads current (I).

EMF electromotive force

EMI electromagnetic interference

EMT electrical metallic tubing. (thin-wall conduit)

EPA Environmental Protection
Agency

Earth terminology for zero reference
ground.

Eddy current induced current flowing
in a magnetic core that is created
by a varying magnetic field.

Edison base fuses 15, 20, and 30 amp
fuses that have the same style
base as an incandescent bulb.

Effective voltage a value of an AC sine
wave voltage that has the same
heating effect as an equal value of
DC voltage. $E_{eff} = E_{peak} \times 0.707$.

Electrical charge the two types of
electrical charges are positive and
negative. Like charges repel and
unlike charges attract.

Electrical degree one 360th of an
alternating current or voltage
cycle.

Electricity energy that produces a flow
of electrons from one atom to an
other.

Electromagnet a magnet made by
passing current through a coil of
wire wound around an iron core.

Electron the part of an atom that carries
a negative charge.

End bell the plate at the end of a motor
that supports the bearings. Also
called end shields or end plates.

Energize to supply power to an
electrical load.

Energy efficiency ratio (EER) the ratio
of capacity in BTU per hour
divided by the watts of electricity
used.

F

F   abbreviation for farad, frequency, fluorine or Fahrenheit.

FLA   full load amps.

Fu   fuse

Factual diagram   a wiring diagram that is a combination of pictorial and schematic diagrams.

Farad   a unit of electrical capacity. When a capacitor is charged with one coulomb of electricity and gives a difference of one volt of potential.

Fast acting fuse   a fuse that opens quickly on overloads and short circuits, not designed for temporary overloads that occur with inductive or capacitive loads.

Field an electrical or magnetic area of force.

Field pole the part of the stator of a motor that concentrates the magnetic field of the field winding.

Fish tape a flexible wire used to pull wires through conduit.

Flux, magnetic the lines of force that connect the north and south poles of a magnet.

Forward bias voltage applied to a P-N junction diode so that the potential barrier is neutralized. (positive voltage on a P-region or negative voltage on an N-region)

Fractional horsepower a horsepower value less than one.

Free electrons electrons in the outer
orbit of an atom that are easily
removed and result in electrical
current flow.

Frequency the number of cycles that an
AC current completes in one
second, expressed in hertz (Hz).

Full load amps current drawn by a
motor when it is operating at
rated load, voltage and frequency.

Fuse an electrical safety device
consisting of a metal strip that
melts when subjected to high
current.

G

GFIC ground fault interrupter circuit.

GRD ground.

Generator a rotating electric machine
that converts mechanical energy
to electrical energy.

Greenfield a flexible metal conduit
used in applications that require
bends at various angles.

Ground an electrical term meaning to
connect a circuit to the earth thus
making a complete circuit. A
common point of zero potential
such as a motor frame or chassis.

H

HP   horsepower

HVAC heating, ventilating, and air conditioning

Hz   hertz.

Hard start kit a kit consisting of a start relay and a capacitor, or a PTC thermistor, to provide high starting torque.

Hertz (Hz) a unit of frequency equal to one cycle per second.

High-voltage circuit a circuit involving a potential of more than 600 volts.

Hot wire a lead with a voltage difference between it and another hot wire or a neutral wire.

I

I   symbol used to designate current.

IBEW International Brotherhood of Electrical Workers.

IC   integrated circuit.

ICE   a device used to help remember that current (I) in a capacitive (C) circuit leads voltage (V).

IMP impedance.

I^2R formula for power in watts.

Impedance the total resistance to the flow of AC current as the result of resistance, reactance, and / or capacitance. It is represented by the symbol Z.

Induced current current that flows as the result of an induced voltage.

Induced voltage the potential that causes current to flow in a conductor that passes through a magnetic field.

Inductance the property of a circuit whereby energy can be stored in a magnetic field .

Inductive circuit any circuit that contains at least one magnetic field.

Insulation materials with few free electrons used to cover wires to prevent short circuits and shock hazards.

Integrated circuit a circuit that uses multiple semiconductors and transistors in a single circuit, sometimes called a chip.

Ion an atom that has gained or lost electrons, resulting in a positive or negative charge.

J

J box a plastic or metal box where
      electrical connections are made.

Jacket the outside cover of a wire or
      cable.

Jogging a quickly repeated switching
      on and off of a motor.

Joule a unit of heat equal to one watt
      second. It is the amount of heat
      needed to raise the temperature of
      one kilogram of water 1°C.

Jumper to connect a wire across the
      contacts of a component for test
      purposes.

Junction a point in an electrical circuit
      where the current branches out
      into other sections.

Junction box see J box

K

Kv   kilovolt.

Kva   kilovolt amps.

Kvah   kilovolt amp hour.

Kw   kilowatt.

Kwh   kilowatt hour.

Kilo-   prefix meaning one thousand.

Kirchoff's current law   the sum of the currents flowing into any point or junction of a circuit is equal to the sum of the currents flowing away from that point.

Kirchoff's voltage law   in any current loop of a circuit, the sum of the voltage drops is equal to the voltage supplied to that loop.

L

L   the symbol for inductance.

L1   incoming power supply line,
        usually 120 volts.

L1-L2 incoming power supply, usually
        240 volts.

L1, L2, L3 incoming 3 phase power
        supply lines.

Ladder diagram a circuit diagram
        drawn in the form of a vertical
        ladder

Law of magnetism the law that states
        that unlike poles of a magnet
        attract and like poles repel.

Lenz's law the induced counter
        electromotive force in a circuit
        will always be in a direction that
        opposes the voltage that produces
        it.

Line a term for the conductors carrying the power from the generating source.

Load center a point from which branch circuits originate.

Locked rotor amps the steady state current drawn by a motor when the rotor is locked to prevent its movement.

Lockout (safety) the opening and locking of the main power switch to safely perform service procedures.

Lugs terminals on the end of a wire or places on equipment to facilitate connections.

M

MA   milliampere -one thousandth of an ampere.

Meg- (or mega-) prefix meaning one million.

mfd   microfarad.

Mho   unit of electrical conductivity of a body with a resistance of one ohm. (The reciprocal of ohm.)

mv   millivolt one thousandth of a volt.

Magnetic field the lines of force that extend from a north polarity and return to a south polarity to form a loop around a magnet.

Main the primary circuit that supplies all other circuits.

Make to complete a circuit by closing a switch.

Megohm resistance equal to 1 million
  ohms.

Micro- prefix meaning one millionth
  part of a specified unit.

Microfarad a unit of capacitance equal
  to one millionth of a farad.

Milli– prefix meaning one thousandth.

Momentary switch a spring loaded
  control that makes or breaks a
  circuit only when it is held in
  place.

Motor a device that changes electrical
  energy to rotating motion.

Multimeter a meter capable of two or
  more electrical quantities, such as
  volts, amps, and ohms.

N

National Electrical Code (NEC)
    sponsored by the National Fire
    Protection Association, a national
    code written for the purpose of
    safeguarding persons and
    property.

Negative charge the charge that results
    from an excess of electrons.

***Negative temperature coefficient
    thermistor (NTC)*** a resistor that
    decreases resistance as
    temperature increases.

Neutral neither positive nor negative
    charge.

Neutron a particle in the center of an
    atom that has no electrical charge.

Nominal the average rating of power
    during normal operation.

Nonferrous the group of metals that
do not contain iron.

Normally closed describes a device that
automatically moves to a closed
position when power is removed.

Normally open describes a device that
automatically moves to an open
position when power is removed.

O

Ohm unit of electrical resistance, its symbol is Ω.

Ohmmeter instrument used to measure resistance.

Ohm's law mathematical relationship, discovered by George Simon Ohm, stated as voltage equals current times resistance.

Open circuit an interrupted electrical circuit that does not provide a complete path for current flow.

Oscillator a device that changes DC voltage into AC voltage.

Oscilloscope a cathode ray tube (CRT) that displays voltage (vertical scale) and time (horizontal scale).

Out-of-phase the condition when two components do not reach their positive and negative peaks at the same time.

Overcurrent a condition in an electrical
circuit when the normal current is
exceeded. Can be caused by an
overload or a short circuit.

P

Pos positive.

Pri primary.

Psc permanent split capacitor.

Ptc positive temperature coefficient .

Panel box electrical junction box that
contains fuses or breakers.

Parallel circuit a circuit that feeds
identical voltage to all branch
circuits or components, with
amperage dividing among the
components.

Peak load the maximum load carried by a unit during a designated period of time.

Peak-to-peak voltage the measurement of voltage from the positive peak to the negative peak of an AC sine wave.

Permanent split capacitor motor a Single phase motor that has a capacitor continuously in the start winding.

Phase angle the degrees of difference between two AC sine waves.

Pole one set of contacts, such as in a switch or relay.

Polyphase generator a generator that rotates three conducting loops (3 phase).

Polyphase motor a motor that operates on 3 phase current.

Positive charge a charge that has a deficiency of electrons.

Positive temperature coefficient thermistor (PTC) thermistor that increases resistance as temperature increases.

Potential relay a relay with normally closed contacts that is energized (opened) by counter EMF across the start windings.

Primary winding the coil of a transformer to which source voltage is applied.

Proton a positively charged particle in the center of an atom.

Q

Quick-connect a solderless terminal with push-on connection.

R

Rf   radio frequency.

Rms   root mean square.

Rpm   revolutions per minute.

Raceway a channel, conduit, or runway for conductors or cables.

Reactance the combined effects of capacitance and inductance on an alternating current, its symbol is "X" and is expressed in units of ohms.

Rectification the process of converting AC into DC.

Relay an electromechanical device that can be energized with a relatively small current to operate a set of contacts that carry a larger current.

Resistance the opposition to current flow.

Rheostat a variable resistor that can be adjusted to various levels.

Root mean square(RMS) the effective value of an alternating periodic current or voltage, calculated as the square root of the average of the squares of all of the instantaneous values of the current (or voltage) throughout one cycle.

Rotor the rotating part of a motor or generator.

Run winding the motor winding that carries current during normal operation.

S

Safety ground the conductor (usually a green or bare wire) that connects the equipment frame or chassis to earth ground.

Safety motor control a device operated by pressure, temperature, or current that will open the circuit if safe conditions are exceeded.

Scale a measurement band on a test instrument.

Secondary voltage the output voltage of a transformer. Can be higher or lower than the primary voltage.

Series circuit a circuit arranged with only one path for current flow.

Service entrance the point of entry from the main electrical power line into the building.

Short circuit an unintentional connection of low resistance resulting in excessive and often damaging current flow between two points of a circuit.

Single phase a device that uses or produces only one alternating current.

Slip the difference between the speed of the rotating magnetic field of a motor and the actual speed of the rotor.

Solenoid an electromechanical device that will move an iron core when energized.

Start winding the winding energized for a brief period to start a single phase electric motor.

Switch a device that makes or breaks contacts to either complete or open a circuit.

T

Tagout the practice of labeling switches to inform others that repairs are in progress.

Terminal a connection point on an electrical device.

Three phase current a combination of three alternating currents that are 120 degrees different in phase.

Torque twisting or rotating force, measured in lbs/ft.

Transformer a device with two or more electromagnetic coils used to increase or decrease AC voltage.

Transistor a semiconductor used to perform switching or amplifying of electrical signals.

U

UL Underwriters Laboratories, an independent testing agency for electrical appliances

Utility transformer a transformer that steps down the utility supply voltage for use in a facility.

V

V volt.

VA volt-ampere.

VAC volts alternating current.

VDC volts direct current.

VOM volt-ohm-milliammeter.

V_{MAX} the maximum voltage in an AC cycle.

V_{RMS} the average voltage (root mean square) in a circuit, equal to the peak voltage time 0.707.

Valance electrons the number of electrons in the outer orbit of the atom determine whether the material is a conductor, insulator or semiconductor.

Volt unit of electromotive force.

Volt-ampere the unit of electrical power.

Volt-ohm-milliammeter (VOM) a meter with multiple functions and ranges, usually including voltage, current and resistance.

Voltage drop the amount of voltage loss from the source through a conductor or load.

W

Watt   the unit of electrical power,
the symbol is W.

Watt's law   in a DC circuit or in a purely
resistive AC circuit, watts equals
volts time amps.

Waveform   a graphic display of voltage
values over a period of time.

Wavelength   the distance between two
corresponding points of two
successive waves of a periodic
waveform.

Wire   a conductor, bare or insulated.

Wire gauge   the system of wire sizing by
diameter, #0000 the largest, #40
and above for the smallest.

Wye connection   also called a star
connection, is made by joining
one end of each of three
windings.

X,Y,Z

X   the symbol for reactance.

Y-axis the vertical axis on a graph.

Z   the symbol for impedance.

Zener diode a diode that has a constant
        voltage drop when operated in the
        reverse direction, often used as a
        voltage regulator.

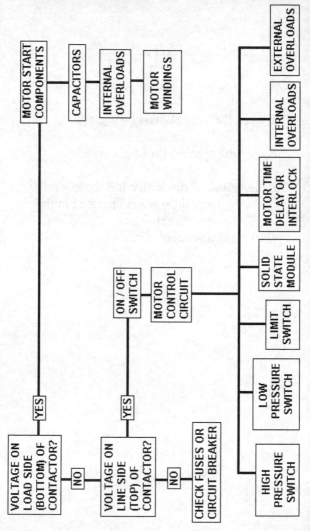

ELECTRICAL FORMULA

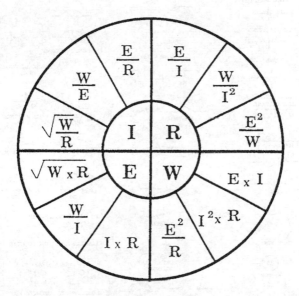

I = Intensity (Amperes)
R = Resistance (Ohms)
E = Electromotive Force (Voltage)
W = Watts

Some Electrical Symbols Used On Schematics

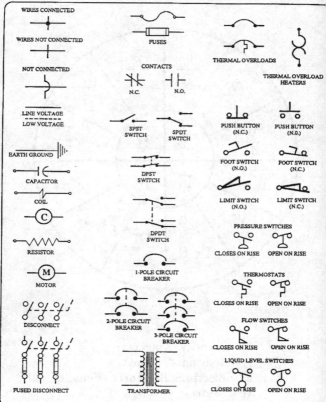

WIRES CONNECTED

WIRES NOT CONNECTED

NOT CONNECTED

LINE VOLTAGE
LOW VOLTAGE

EARTH GROUND

CAPACITOR

COIL

RESISTOR

MOTOR

DISCONNECT

FUSED DISCONNECT

FUSES

CONTACTS

N.C.   N.O.

SPST SWITCH   SPDT SWITCH

DPST SWITCH

DPDT SWITCH

1-POLE CIRCUIT BREAKER

2-POLE CIRCUIT BREAKER   3-POLE CIRCUIT BREAKER

TRANSFORMER

THERMAL OVERLOADS

THERMAL OVERLOAD HEATERS

PUSH BUTTON (N.C.)   PUSH BUTTON (N.O.)

FOOT SWITCH (N.O.)   FOOT SWITCH (N.C.)

LIMIT SWITCH (N.O.)   LIMIT SWITCH (N.C.)

PRESSURE SWITCHES

CLOSES ON RISE   OPEN ON RISE

THERMOSTATS

CLOSES ON RISE   OPEN ON RISE

FLOW SWITCHES

CLOSES ON RISE   OPEN ON RISE

LIQUID LEVEL SWITCHES

CLOSES ON RISE   OPEN ON RISE

CONDUCTANCE AND RESISTANCE OF VARIOUS METALS		
METAL	**CONDUCTANCE**	**RESISTANCE**
Silver	2207	1.0
Copper	2030	1.09
Gold	1471	1.50
Aluminum	1277	1.73
Tungsten	634	3.48
Brass	461	4.79
Iron	304	7.26
Nickel	263	8.39
Steel	235	9.39
Nichrome	34	64.9l

MOTOR SPEEDS		
NUMBER OF POLES	**SYNCHRONOUS SPEED**	**ACTUAL SPEED**
Two-pole Motor	7200 ÷ 2 = 3600 RPM	3450 RPM
Four-pole Motor	7200 ÷ 4 = 1800 RPM	1725 RPM
Six-pole Motor	7200 ÷ 6 = 1200 RPM	1150 RPM

INSULATED CONDUCTOR APPLICATIONS			
TYPE OF INSULATION	LETTER CODE	MAX. TEMP.	APPLICATION
Asbestos	A	392F	Dry locations only
Asbestos and Varnished Cambric	AVA	230F	Dry locations only
Heat Resistant Rubber	RH	167F	Dry and damp locations
Thermoplastic	T	140F	Dry locations
Moisture Resistant Thermoplastic	TW	140F	Dry and wet locations
Heat Resistant Thermoplastic	THHN	194F	Dry locations
Moisture and Heat Resistant Thermoplastic	THW or THWN	167F	Dry and wet locations
Underground Feeder	UF	140F	Moisture resistant
Varnished Cambric	V	185F	Dry locations only

TAP	COLOR
COMMON	WHITE
HIGH	BLACK
MEDIUM	YELLOW
LOW	RED
CAPACITOR	PURPLE (2 WIRES)

Table shows colors used by manufacturers to code wires used when making external connections to multi-speed motors.

HP	RUN WINDING	START WINDING
1/8	4.5 Ω	16 Ω
1/6	4.0 Ω	16 Ω
1/5	2.5 Ω	13 Ω
1/4	2.0 Ω	17 Ω

This table shows run and start winding resistance, in ohms, for common sizes of fractional horsepower single-phase.

Coil No.	Pick Up Voltage	Drop Out Voltage	Continuous Voltage	Coil Ohms (approx.)
1	139-153	15-55	130	760
2	140-153	20-45	170	1400
3	159-172	35-77	256	3320
4	261-290	50-100	336	5180
5	280-310	50-100	395	7150
6	299-327	50-100	420	10000
7	323-352	60-135	495	11950

This table shows the pickup voltage ranges and other information on different relay coils that are available.

TO FIND	DIRECT CURRENT DC	ALTERNATING CURRENT AC SINGLE PHASE	ALTERNATING CURRENT AC THREE PHASE
Amperes when horsepower is known	$\dfrac{Hp \times 746}{E \times \%Eff.}$	$\dfrac{Hp \times 746}{E \times \%Eff \times PF.}$	$\dfrac{Hp \times 746}{1.73 \times E \times \%Eff \times PF.}$
Amperes when kilowatts are known	$\dfrac{kW \times 1000}{E}$	$\dfrac{kW \times 1000}{E \times PF}$	$\dfrac{kW \times 1000}{1.73 \times E \times PF}$
Amperes when KVA is known		$\dfrac{KVA \times 1000}{E}$	$\dfrac{KVA \times 1000}{1.73 \times E}$
Kilowatts	$\dfrac{I \times E}{1000}$	$\dfrac{I \times E \times PF}{1000}$	$\dfrac{I \times E \times 1.73 \times PF}{1000}$
KVA		$\dfrac{I \times E}{1000}$	$\dfrac{I \times E \times 1.73}{1000}$
Horsepower	$\dfrac{I \times E \times \%Eff.}{746}$	$\dfrac{I \times E \times \%Eff \times PF}{746}$	$\dfrac{I \times E \times 1.73 \times \%Eff \times PF}{746}$

DIMENSIONS, TYPICAL RESISTANCES AND AMPACITY OF COMMERCIAL WIRE					
		CURRENT CAPACITY (AMPERES)			
		COPPER		ALUMINUM	
GAUGE NO. (AWG)	DIAMETER BARE WIRE (INCHES)	TW UF	RH, RHW, THHW,THW, THWN	TW UF	RH, RHW, THHW, THW, THWN
0000 (4/0)	0.460	195	230	150	180
000 (3/0)	0.410	165	200	130	155
00 (2/0)	0.365	145	175	115	135
0 (1/0)	0.325	125	150	100	120
1	0.289	110	130	85	100
2	0.258	95	115	75	90
3	0.229	85	100	65	75
4	0.204	70	85	55	65
6	0.162	55	65	40	50
8	0.128	40	50	30	40
10	0.102	30*	30*	20	25
12	0.081	20*	20*	16	18
14	0.064	15*	15*		
16	0.051	10*	10*		
18	0.040	5*	5*		